Explorers in
Dinosaur
World

By Geoffrey T. Williams
Illustrated by Robert Cremins

PRICE STERN SLOAN
Los Angeles

Explorers in Dinosaur World can be read with
its accompanying word-for-word audiocassette
to create an exciting, educational adventure
into the past.

Copyright © 1987 by Geoffrey T. Williams
Illustrations copyright © 1987 by Price Stern Sloan, Inc.
Published by Price Stern Sloan, Inc.
360 North La Cienega Boulevard, Los Angeles, California 90048

ISBN: 0-8431-2264-1

10 9 8 7 6 5 4 3 2

PRONOUNCIATION KEY

ANKYLOSAUR	(an·KILE·uh·sor)
APATOSAUR	(ah·PAT·uh·SOR)
ARCHELON	(AR·kee·lon)
BRACHIOSAUR	(BRAK·ee·uh·SOR)
CORYTHOSAUR	(ko·RITH·uh·SOR)
DEINONYCHUS	(dine·ON·ik·us)
DIPLODOCUS	(dih·PLOD·uh·kus)
ELASMOSAUR	(ee·LAZ·mo·sor)
HESPERORNIS	(hes·per·OR·nis)
ICHTHYOSAUR	(IK·thee·uh·SOR)
IGUANODON	(ih·GWAN·uh·don)
KRONOSAUR	(KRON·uh·sor)
PACHYCEPHALOSAURUS	(PAK·ee·SEF·uh·low·SOR·us)
PENTACERATOPS	(pen·ta·SER·uh·tops)
PHOBOSUCHUS	(fo·bo·SOOK·us)
PLESIOSAUR	(PLEE·zee·uh·SOR)
PTERANODON	(ter·AN·uh·don)
PTERODACTYL	(ter·uh·DAK·til)
QUETZALCOATLUS	(ket·sul·ko·AT·lus)
RHAMPHORHYNCHUS	(ram·fo·RINK·us)
SALTOPUS	(SALT·o·pus)
SPINOSAUR	(SPY·nuh·SOR)
TRICERATOPS	(try·SER·ah·tops)
TYRANNOSAURUS REX	(tye·ran·ah·SOR·us rex)

"Technically, ichthyosoars aren't dinosaurs," Pete said drowsily, as he climbed into his sleeping bag.

"You're right, Pete. But her discovery led some scientists to investigate other strange bones and fossils, such as the dinosaur bone Reverend Robert Plot described back in 1677. He said it was the bone of a giant human." Pete chuckled at this. Then, as the moon made its way through the constellations, shining like a distant beacon on the primeval forest, Jake told them other stories. Such as the rivalry between paleontologists Othniel Charles Marsh and Edward Drinker Cope. "Back in the 1800's they each sent teams out West, into Colorado and Wyoming, to find dinosaur fossils. There was a dinosaur fever in the country. People just couldn't get enough. Anyway, Marsh and Cope were so jealous that they hid their discoveries from each other. And, of course, the West was really wild back then so they had to fight off Indians…" At about this point in his story, the kids dropped off to sleep. Pete had a smile on his face, dreaming about Indian attacks and famous scientists and huge dinosaur bones never seen before. Wendy dreamed of a little girl making the fossil discovery of the century and then going on to become a world-famous paleontologist. In her dream the little girl looked just like Wendy.

They woke up the next morning to the smell of hot chocolate. Mmm-mm! Jake also fixed pancakes and bacon and, since most explorers would tell you nothing makes you hungrier than the prospect of an exciting adventure, it wasn't long before the three of them were wiping their plates clean and getting ready to leave. The kids did the dishes while Jake packed and cleaned up their campsite. Suddenly, a voice interrupted their morning chores. "Pangaea Camp One this is Main Base. Main Base calling Pangaea Camp One. Come in, Jake."

Jake picked up the microphone on the portable radio, "Main Base this is Camp One. Go ahead."

"Security reports a break in the perimeter fencing on the east side of the lake."

"How bad is it, Main Base?"

"Bad enough. Tracking computers indicate a pack of deinonychus is on the loose, and Jake…they're heading your way. Suggest you make for the chopper pad. We have a bird on its way now. You should have plenty of time to make it. Do you copy?"

"We're on our way. This is Pangaea Camp One over and out." Jake was all business now. "You heard him. Let's go, kids."

The trail to the helicopter landing pad led past a swamp where several gigantic phobosuchus were floating lazily. As the trio passed, the reptiles hissed menacingly, their cavernous jaws snapping open and closed, showing off rows of long, needlelike teeth. To Pete, they looked just like crocodiles—fifty-foot-long crocodiles. To Wendy they just looked frightening.

Soon the path turned away from the lake and ran beneath a leafy tunnel of oak trees. The light was dim and murky. Suddenly, Pete stopped and looked back down the path, "Jake, there's something behind…"

But before he could finish, there was a horrendous crashing and cracking of branches and underbrush, and a huge creature lumbered onto the path directly in front of them.

It was bigger than an army tank and almost as heavily armored. Its back and sides were covered with bony plates set close together in thick, leathery skin. It had a row of stubby spikes that protected each side of its massive body. But its most unusual feature was its tail: short and thick and ending in a big, knobby ball, which it was swinging back and forth like a club.

"An ankylosaur. And just in time," Pete whispered in awe.

"What do you mean, 'just in time'?" Jake asked.

But there was no time to answer as they heard the thunder of clawed feet pounding up the path behind them.

Ankylosaurs in front, deinonychus in back! Jake looked for an escape. "Quick, through here!" They darted between a break in the line of trees just as the snarling pack of dinosaurs appeared and caught sight of them. If you've ever seen a deinonychus, you know how vicious they are. The sharp talons on their hands are perfectly made for clutching, while the hooked, razor-sharp claws on their feet rip open their helpless prey. The savage pack was ravenously hungry and mean-tempered and nothing that stood between them and their intended victims was safe. Nothing except an ankylosaur.

This was one plant-eater that could take care of itself. When the carnivores attacked, the ankylosaur's prodigious tail went into action, whipping back and forth, sending the hapless deinonychus sprawling in all directions. Those that did get close enough found their sharp claws had no effect on the creature's armored hide. The battle echoed through the forest. Roaring, screaming, howling, hissing, tearing, biting. And through it all, the devastating, bone-crunching THUMP of the ankylosaur's tail. Soon it was over. The deinonychus scattered, limping away in defeat, and the ankylosaur resumed its peaceful walk along the path. And the explorers had seen it all.

The helicopter lifted off toward the mainland, its blades whipping up dust and leaves. As they rose above the tangled forest the island of Pangaea was spread out before them in a sweeping panorama—a timeless world of wonder and danger, beauty and surprise.

Jake was looking thoughtfully at Pete. "You knew what would happen when the deinonychus met the ankylosaur, didn't you?"

"It was just a lucky guess," Pete answered modestly.

"Lucky for us, Pete." Then Jake turned to Wendy, "Do you think your brother would be interested in a job here during summer vacation?"

Pete was grinning so hard Wendy thought his face would break. "Gee, Jake, I don't know if he'll have time. He has dishes to do. And setting the table. And math homework…"

"Wendy…" Pete growled. And both of the young explorers broke up in laughter.

If there's anything you want to know about dinosaurs, ask Peter Wynne. If there was an animal that had teeth longer than butcher knives, a mouth the size of your front door and made footprints you could park your dad's car in, chances are Pete knew its name and how to spell it, where it lived, how fast it ran and what it wanted for dinner.

In his room were pictures of a tyrannosaurus rex fighting a huge triceratops, green iguanadons browsing next to giant brown brachiosaurs and two pachycephalosaurus ramming one another with their massive heads; a wooden pterodactyl floated silently on the end of a string above his bed; on his bookshelf were painted models of a diplodocus, a spinosaur and a snake-necked, fish-finned plesiosaur; on one wall was a map of his all-time favorite place—DINOSAUR WORLD. Pete thought his room was wonderful. His mother thought it was a swamp.

Pete was sitting at his desk painting the wickedly sharp claws on a model of a snarling deinonychus, when suddenly the door burst ope and his little sister, Wendy, stuck her head in and yelled, "You're gonn miss the drawing!" Then she slammed the door. The brush in Pete's hand slipped, sending a glob of white paint spattering across the fac of one of the meanest-tempered dinosaurs that ever lived.

The boy called menacingly, "Wendy!" But it was too late. She was gone. Then he looked at his watch and was startled to realize that his sister was right—it *was* time for the drawing! If he didn't hurry, he'd miss it. He turned on the radio just in time to hear the jingle sing, "W-DINO." Then the disc jockey said, "Hey! You're never too old to rock 'n roll, 'cause the hits are never extinct!" There was the sound of a drum roll, and the DJ announced in a deep, echoing voice, "And now, the moment you've all been waiting for…"

Pete turned the radio up.

"That's right! The W-DINO big dinosaur drawing! Our lucky winner and his or her special guest will go where no outsider has gone before! Dinosaur World's newest attraction—Pangaea—the island of mystery in the middle of Dinosaur Sea!"

"I hope, I hope, I hope it's me," Pete was saying over and over. Every kid he knew had sent cards in, including Wendy. But he had sent in dozens and dozens—more than anyone. He was going to win—he just knew it.

The DJ said, "Our winners will sail across Dinosaur Sea to Pangaea where they'll spend an entire weekend exploring the tangled jungles and glittering beaches. Now…I'm reaching into the box for the winning card…I've got it…Yes! I'm pulling it out. I'm reading the name of our lucky winner…"

Wow! A whole weekend in Dinosaur World! Pete was holding his breath.

"It's…" the DJ paused dramatically.

"It's me! It's me!" Pete yelled at the radio.

"It's Wendy Wynne!"

"Oh no," sighed Pete as he heard a whoop of joy from down the hall. "It's my sister."

Mornings in Dinosaur World usually start out quietly. Little hisses and squeaks come from the Nursery where baby dinosaurs wake up, crying for something to eat; there are distant squeals from the pteranodons in their cliff-top nests; in the forest the ringing call of a big duckbill can be heard. But this Saturday morning was different.

There was already a line of people waiting to buy tickets when a long, low limousine pulled up to the entrance. The crowd buzzed with excitement. As the uniformed chauffeur opened the back door people stood on their toes trying to see over each other. Who was in the car?

It wasn't a movie star. It wasn't a rock group. And it wasn't the President. It was a little girl. Someone whispered, "It's her!"

Another voice yelled, "It's Wendy Wynne. The girl who won the drawing. Way to go, Wendy!"

Wendy smiled a little self-consciously.

A man walked up to the limousine. He was dressed in boots, blue pants, a khaki shirt and tie and a lightweight jacket with a Dinosaur World patch on the sleeve. He smiled at the girl, "Welcome to Dinosaur World. My name's Jake DuMel. I'm your…" He was interrupted by a piercing shriek. There was a loud flapping sound and a rush of wind as a huge shadow soared over the crowd. A few people screamed and ducked down. "The pterodactyls seem to be flying a little low today," Jake observed. "As I was saying, welcome to Dinosaur World. I'll be your guide to Pangaea."

Wendy was still looking at the enormous creature in astonishment, "I thought pterodactyls were little. Like…like ducks or chickens," she said.

Just then, a voice came from inside the limousine, "Some pterodactyls, namely the quetzalcoatlus of the late Cretaceous, had wingspans of over forty feet. You're probably thinking of the little model hanging up in my room." As you've probably guessed by now, it was Peter.

Not winning the drawing had disappointed him, but he'd finally convinced his sister to bring him along as her guest. It hadn't taken much. Just a promise to set the table, wash the dishes and help her with her math homework every night for the next year, that's all.

"This is my brother, Pete. He knows everything about dinosaurs," she explained matter-of-factly as they unloaded their camping gear.

Jake smiled, "Well, he sure knows about pterodactyls. Except they aren't really dinosaurs. They're flying reptiles."

"I know. Pete told me," she boasted.

Jake smiled. It was obvious Wendy adored her brother.

On the way to the boat dock they walked through Mesozoic Mesa past the Petting Zoo, the snack stands and the Nursery. People dressed in neat Dinosaur World uniforms were hurrying this way and that. "What do they all do?" Wendy asked.

"It takes a lot of people to run a place this big. Engineers for the T-Rex Express trains, guides for group tours—teachers are always bringing their classes here—keepers who help in the Nursey feeding the baby dinosaurs, gardeners to take care of all the plants and flowers and a staff of scientists who spend time studying the animals."

They passed a herd of about twenty duckbill dinosaurs. "Corythosaurs," Pete informed his sister about the slow-moving animals that were grazing in a grove of pine trees. The largest one was over thirty feet long and must have weighed as much as two or three elephants. Their skin was pebbly, like the surface of a football, and colored in green and brown patches that helped them blend in with the forest and hide from hungry meat-eating dinosaurs. Pete knew they were corythosaurs because of the rounded, helmet-shaped crests, like weird rooster combs, on their heads. They were making honking sounds as they ate. Two men with clipboards were standing nearby taking notes. Pete asked Jake what they were studying.

"They're monitoring the corythosaurs' diet to see if we need to move some of them to an area where there's more food."

"How much do they eat?" asked Wendy.

Jake said, "That big guy probably needs two hundred pounds of food a day. In fact, they need so much food, eating is about all they have time for."

Almost as though it had heard him, the giant duckbill lifted his head, snorted and then gave a trumpeting blare that made Wendy cover her ears.

Pangaea was about a half a mile from the mainland. Pockets of early morning fog still nestled on Dinosaur Sea, making the island appear ghostly and mysterious. Smoke curled from several small volcanos. Flying creatures, too far away to identify, circled the island's rocky summit. A stray breeze scattered wisps of fog and Pete could see tall, reedy plants clustered thick in a swamp and strange, broad-leafed trees with gnarled and twisted roots growing at the base of towering cliffs.

Occasionally the smooth surface of the water rippled as though unseen creatures swam just below. Pete thought he saw something that could have been a long neck arch up and out of the water in the murky distance, but when he blinked it was gone. As he turned to say something to Jake he caught Wendy's eye and knew she'd seen it too. Probably just a tree branch, he thought.

Jake was busy checking their backpacks and gear. "Sleeping bags. Cooking utensils. Flashlights. Food. Radio…"

"Why do we need a radio? Do dinosaurs listen to W-DINO too?" Wendy wanted to know.

"It's not that kind of radio," her brother said. "We need it to call for help if we get into any trouble."

"Dinosaur World is full of surprises, not trouble. Still, you can't be too careful," Jake assured the little girl. Then he loaded everything into their small boat, the *Intrepid,* and helped them aboard. After they had all put on lifejackets, he started the engine and cast the line off from the dock. As they set off across the water to the dim, fog-shrouded island in the distance, Pete grinned and said, "Now the adventure begins!"

A small flock of gulls, hoping for a stray bite of food, followed behind *Intrepid*.

Other things followed the boat as well—creatures Pete had never seen before—almost like birds, about four feet long, with narrow sharp beaks that they clicked together, making a sound like drumsticks on a tabletop.

"Those are hesperornis," Jake told him. "They're primitive birds. They remind people of loons. They can't fly at all but they're great swimmers and divers." As they watched, several of the birds squawked and disappeared underwater, emerging moments later clutching wriggling fish in their sharp teeth.

"What's that behind us?" Wendy asked. She pointed to a long, hump-backed shape that was moving slowly towards them. It looked like a boat turned upside-down. Except it was very much alive and bigger than any boat in Dinosaur World.

Jake immediately increased the *Intrepid's* speed until they were further away. "That's something we don't want to tangle with."

"A plesiosaur," said Pete, almost in a whisper.

"The biggest plesiosaur that ever lived: an elasmosaur. Almost sixty feet long. And there's only one thing it likes to eat as much as fish…"

The sea monster's head emerged from the water. Pete gasped and Wendy gave an astonished cry. Its neck was over twenty feet long! It arched up and up until it towered over the boat like an immense snake. It opened its mouth and hissed. Water dripped from its gaping jaws and they could see its deadly teeth glittering. Its bright eyes immediately spotted a hesperornis desperately trying to escape. Like a huge whip, the elasmosaur's head lashed out and in the blink of an eye there was nothing left of the unfortunate bird but a few feathers floating on the water. Then the immense reptile turned its unblinking gaze toward the people in the little boat, eyeing them curiously. But after a moment, its appetite apparently satisfied, it turned and slowly swam away.

Pete said "Wow!" in an awed voice, while Wendy let out her breath in a relieved sigh.

Jake said, "It's a good thing our boat goes faster than it does or we might be in trouble." At that moment the *Intrepid's* engine coughed once and died.

The small boat rose and fell on the waves. Without power it was at the mercy of the sea. In the quiet, Wendy seemed to hear ordinary sounds louder than they actually were: water lapping against the hull, the faraway rumble of volcanos, the mew and cry of a gull overhead. She looked at her brother. He was standing up in the front of the boat, shading his eyes and gazing across the water like an explorer in a new land. He turned and grinned at her and she felt better.

"Hmmmm," said their guide. "Guess it's time for me to earn my keep." With that, Jake pulled two oars from the bottom of the boat, passed them through the oarlocks and began rowing toward Pangaea.

They were still several hundred yards from shore when Wendy yelled, "Look!"

The colossal creature churning through the water towards them was perhaps the most terrible monster to ever hunt in the ancient oceans of the earth: a kronosaur. It was over fifty feet long and had jaws even bigger than the great tyrannosaurus rex—over eight feet long. As it swam, its massive tail whipped the water to foam.

Just then, Pete saw another sea-giant nearby—an enormous turtle bigger than a car—and it was heading straight for Pangaea! It gave the boy an idea. He pawed furiously through his pack until he came up with a long rope. "Jake! If we could lasso that archelon maybe it could tow us to shore!"

"Good thinking, Pete!" The kronosaur was obviously after the ar- chelon, but their boat was in the way. Jake made a loop and whirled it over his head, just like a cowboy roping a steer, only he was trying to rope the biggest turtle that ever lived. The rope spun silently across the water, landing perfectly over the archelon's head. The determined beast never even noticed—it just kept swimming, pulling the boat and its passengers out of the jaws of danger to the safety of the island. Behind them the hungry kronosaur roared in frustration and anger as it watched its breakfast escape.

The fog was clearing and the sun felt warm on their backs as Jake, Pete and Wendy pulled the boat up onto the sand. Now that she was safe on land again, Wendy found herself feeling more cheerful and looking forward to whatever the next moment would bring. As for Pete, the adventure was even better than he had imagined. Dinosaur World was thrilling and mysterious. The very air seemed charged with excitement.

Looking around, they found themselves in a world impossibly old and, at the same time, wondrously new. Jagged cliffs, the ancient bones of the earth, towered hundreds of feet over their heads. At the swamp's edge stood a forest of giant ferns, looking much as they had at the very dawn of the Permian age, millions of years before even the dinosaurs walked the earth. Wading in the muddy water was a colossal apatosaur, bigger than a moving van, from the Late Jurassic age, which existed over 125 million years ago. And yet, Pangaea was brand-new—they were the first visitors to leave foot-prints on its sandy shoreline, and here was the *Intrepid,* evidence of high (although broken) technology, and, across the sea on the mainland, were highways and cars and electricity and television.

As they shouldered their packs a tiny dinosaur, no bigger than a cat, scampered by. It was running on its hind legs, bent over, balancing itself with its long tail. Its hands, each with five tiny fingers, were outstretched, trying to catch an elusive lizard. It was running so fast that its whole body, about two feet long, was bouncing from side to side. Wendy couldn't help but laugh, "It's so funny! What is it?"

"It's a saltopus," her brother said.

And Jake added, "I'll bet that lizard doesn't find it funny."

With that, the tiny dinosaur snatched up the lizard and made a quick meal of it.

At the base of the cliff behind them was a smooth path which led up in easy switchbacks for several hundred feet before turning toward the middle of the island. The trio paused for one more look at Dinosaur Sea. Bright sunlight danced on the waves, giving the water a cheerful and inviting look—but they knew what perils waited just below the surface of that seemingly peaceful ocean.

From somewhere ahead, deep in Pangaea's unexplored interior, came a loud roar and Pete wondered what adventure the next turn in the trail would bring.

The path to the interior of the island wound between two large granite outcroppings and, as the explorers turned a corner, Dinosaur World revealed yet another incredible surprise, for surrounded by the rugged cliffs was the breathtakingly beautiful Hidden Valley of Pangaea.

Nestled within the valley was a lush Mesozoic forest, as green and rich as a treasure chest of emeralds. In the distance, several small rhamphorhynchus circled lazily in the warm updrafts; bright birds fluttered from tree to tree, filling the air with their songs; larger shapes, too far away to see clearly, moved in the shadows of the forest. At their feet, a crystal clear spring bubbled up from the ground, forming a narrow stream that twisted and tumbled its way down through the forest to the brilliant blue waters of a small lake. Jake, who had helped design and build the island, stood quietly. Pete and Wendy gazed across the wonderland that was Pangaea.

Pete was the first to speak, "Where do we go from here?"

"I thought we'd camp out close to Richard Owen Lake," answered Jake.

"Who was Richard Owen?" Wendy asked.

"He was the British paleontologist who made up the name dinosaur," Jake answered.

And Pete added, "Dinosaur means 'terrible lizard'."

Jake nodded, "And speaking of terrible lizards, there's one on Pangaea we have to be on the lookout for—deinonychus."

"Are they big and ferocious like tyrannosaurus rex?" asked Wendy, who couldn't imagine anything more frightful.

"No. They're quite a bit smaller. Maybe ten feet long. But they run like the wind, hunt in packs like wolves, are always hungry and have extremely long, sharp claws. In fact, deinonychus means 'terrible claw'."

"Can I go home now?" asked the little girl.

Jake chuckled. "Don't worry. You can't see them from here, but there are special fences that keep dangerous dinosaurs from wandering into the areas where we'll be going."

Late that afternoon they reached Richard Owen Lake. They set up camp in a shaded grove of flowering yews and magnolia trees close to the water's edge and soon Jake had a small fire going.

Wendy and Pete went for a walk along the edge of the lake. Reeds poked out of the shallow water near the shoreline. Hundreds of frogs made a weird chorus, their voices twanging like rubber bands. A flock of ducks cruised by. A heron, perched on one stick-thin leg, eyed the children curiously, perhaps wondering what kind of strange new creatures these were. As they walked around a large boulder they came face-to-face with two monstrous dinosaurs drinking from the lake. Wendy was startled and immediately reached for her brother's hand.

Pete whispered, "It's all right, Sis. They're just pentaceratops. Like giant cows—not dangerous at all!"

To Wendy's eyes these behemoths were bigger than any five cows, but she relaxed a bit and studied them, still holding tightly to Pete's

hand. She knew enough about dinosaurs to know that pentaceratops looked like triceratops: their necks were covered by a bony,
fan-shaped frill, they had two long sharp horns just above their eyes
and a shorter nose horn. What made the pentaceratops different
were two additional hornlike growths, one on each cheek.

"Usually they travel in giant herds," Pete said. "And the big meat-
eaters pretty much stay away. Those horns could hurt even an
allosaur." One of the animals looked up and snorted, blinking its big,
dark eyes at the strangers. But Pete and Wendy didn't look particu-
larly menacing, so the giant herbivore shook its massive head back
and forth and went back to drinking. When they were done, they
lumbered ponderously out of the water and through the thick forest
undergrowth, leaving deep footprints in the soft earth.

No visitor had ever spent the night in Dinosaur World before. For Pete and Wendy, this was a magic time. A time for talking about the day's adventures, telling stories and listening to the sounds of the deepening night. The sun had set, a shimmering ball of fire that turned the sky red and orange. Crickets were busily tuning up for their evening performance. Frogs croaked, birds fluttered in the trees and larger, stranger animals, veiled by the darkness, growled and pushed their way through the underbrush.

The moon rose like a silver ornament over the jagged cliffs, silhouetting a large pteranodon winging its way toward its cliff-top nest. Eerie shadows moved through the forest. Dark trees swayed in the soft breeze. Midnight shapes stirred, blurred insubstantial things that could have been vines—or tails. Was the firelight reflecting off luminous eyes? Or was that just a tired child's imagination?

"Wendy, did you know it was a little girl about your age who discovered one of the first dinosaurs?" Jake asked.

Wendy covered a yawn and asked, "Really? A little girl? Who was she

"Her name was Mary Anning. And she found the complete fossil of an ichthyosaur—like a prehistoric dolphin—imbedded in the side of a cliff near her home in southern England."